Smart Cities and Towns

Heather Hammonds

Smart Cities and Towns

Text: Heather Hammonds
Publishers: Tania Mazzeo and Eliza Webb
Series consultant: Amanda Sutera
Hands on Heads Consulting
Editor: Kirstie Innes-Will
Project editor: Annabel Smith
Designer: Leigh Ashforth
Project designer: Danielle Maccarone
Permissions researcher: Lumina Datamatics
Production controller: Renee Tome

Acknowledgements
We would like to thank the following for permission to reproduce copyright material:

Front cover: Fahroni/Alamy Stock Photo; p. 4: Jon Arnold Images Ltd/Alamy Stock Photo; p. 5: (top) Sean Pavone/Shutterstock.com; (bottom) Don Mammoser/Shutterstock.com; p. 6: Johnstocker Production/Shutterstock.com; p. 7: (top left) Gorynvd/Shutterstock.com; (top second left) MikeDotta/Shutterstock.com; (top third left, p. 14) Alexey Panferov/Alamy Stock Photo; (top right) Gorodenkoff/Shutterstock.com; (bottom) Andrej Sokolow/picture alliance/Getty Images; p. 8: (top) nazar_ab/E+/Getty Images; (bottom) CarlosBarquero/Shutterstock.com; p. 9: (top) VesnaArt/Shutterstock.com; (bottom) Dragana Gordic/Shutterstock.com; p. 10: Sombat Muycheen/Shutterstock.com; p. 11: (top) Sombat Muycheen/Shutterstock.com; (bottom) wedninth/Shutterstock.com; p. 12: Leonid Andronov/Shutterstock.com; p. 13: Sergey Dzyuba/Shutterstock.com; p. 15: (top) Summit Art Creations/Shutterstock.com; (bottom) Pandora Pictures/Shutterstock.com; p. 16: (top) anystock/Shutterstock.com; (bottom) YAY Media AS/Alamy Stock Photo; p. 17: (top) (content page) Soonthorn/Adobe Stock Photos; (bottom) Graham King/Shutterstock.com; p. 18: (bottom) SouthWestImagesAus/Shutterstock.com; p. 19: (top) Igor Sorokin/Alamy Stock Photo; (bottom) Marc Dozier/The Image Bank Unreleased/Getty Images; p. 20: (bottom) Richard Whitcombe/Shutterstock.com; p. 21: (top) Igor Sorokin/Alamy Stock Photo; (bottom) (title page) IRINA/Adobe Stock Photos; p. 22: (bottom) Clarence Holmes Photography/Alamy Stock Photo; p. 23: (top) Igor Sorokin/Alamy Stock Photo; (bottom) Marc A. Hermann/MTA/CC BY 2.0; p. 24: (top) (back cover left) Yuichiro Chino/Moment/Getty Images; (bottom) UnknownLatitude Images/Shutterstock.com; p. 25: (top) RelaxFoto.de/E+/Getty Images; (bottom) New Africa/Adobe Stock Photos; p. 26: (left) DOD Photo/Alamy Stock Photo; (right) Quang NGUYEN DUC/Alamy Stock Photo; p. 27: (top left) Lili.Q/Adobe Stock Photos; (top right) (back cover right) georgeclerk/iStock/Getty Images; (bottom) ROSLAN RAHMAN/Getty Images; p. 28: Kiyoshi Hijiki/Alamy Stock Photo; p. 29: (top) Morsa Images/DigitalVision/Getty Images; (bottom) Phonlamai Photo/Shutterstock.com; p. 30: (top) Morsa Images/DigitalVision/Getty Images; (bottom) Oleksandr Rado/Alamy Stock Photo.

Every effort has been made to trace and acknowledge copyright. However, if any infringement has occurred, the publishers tender their apologies and invite the copyright holders to contact them.

NovaStar

ISBN 978 0 17 033501 0

Cengage Learning Australia
Level 5, 80 Dorcas Street
Southbank VIC 3006 Australia
Phone: 1300 790 853
Email: aust.nelsonprimary@cengage.com

For learning solutions, visit **cengage.com.au**

Printed in China by 1010 Printing International Ltd
1 2 3 4 5 6 7 29 28 27 26 25

Nelson acknowledges the Traditional Owners and Custodians of the lands of all First Nations Peoples. We pay respect to Elders past and present, and extend that respect to all First Nations Peoples today.

Many thanks to Richard and Meredith Begg for their assistance with this book.

Contents

Our Changing Cities and Towns

Today, there are more than eight billion people on Earth. It's believed that by the year 2050 there may be around 9.7 billion people on the planet.

All people need places to live, work and play. Many choose to live in large cities or in towns nearby. Living in a city or a town allows people to be close to work, shops, restaurants, parks and other places.

Major cities are very crowded places. Millions of people move around them, day and night.

Cape Town, in South Africa, is a large city with more than 4.5 million people living in it.

Tokyo, in Japan, is currently the largest city in the world, with a population of more than 37 million people.

As the world's cities and towns grow larger, issues such as overcrowding, traffic jams and the safety of all **citizens** are managed by city **authorities**. Governments and businesses around the world use the latest technology to do this. As a result, large cities and towns are becoming more **liveable**.

Protecting the environment is also important in large cities and towns. Many cities are introducing energy-saving measures, and creating green spaces. These measures help to keep the planet healthy, and make cities and towns more **sustainable**.

Delhi, in India, is a megacity with a population of 33.8 million people.

MEGACITIES

Cities with more than 10 million people are sometimes called megacities. Managing overcrowding is a challenge in megacities.

Cities, Towns and Modern Technology

Cities and towns that use modern technology to provide a range of **services** are often called "smart cities" or "smart towns". The technology is used to improve things like communications, traffic and pedestrian management, public transport, electricity supply, water supply and waste removal. Technology can also be used when designing parks and other spaces, such as streets and town squares.

The Internet of Things

The Internet of Things, or "IoT", is the name for a group of devices that are connected by communications **networks**, such as the internet, and are used to collect information. In smart cities and towns, the Internet of Things is thousands of small cameras, **sensors** and other devices which send and receive information over networks.

Devices within the Internet of Things in a city monitor things such as temperature and security.

There are many uses for the information gathered using the Internet of Things. The information helps governments make good decisions. For instance, the information helps them see where car parking spaces are most needed or where to put traffic lights on roads.

In smart cities and towns, information is often transmitted, or sent, immediately, so it can be used as quickly as possible. The information collected is used by government departments and other services. For example, it is used to control traffic.

he Internet of Things in a city is made of many devices connected to the internet.

SMART HOME APPLIANCES

Some smart home appliances are part of the Internet of Things, too! They have small cameras and sensors, and can connect to the internet.

A smart fridge monitors temperature and even helps you keep track of how much food is inside it.

Cities and **Towns** **Connecting People**

Smart cities and towns use technology to help people connect with the communities around them. **Smart devices** such as mobile phones, smart watches and tablets allow people to connect with others quickly and easily.

Everyone with a smart device can use free local Wi-Fi to access the internet. It is readily available in smart cities and towns.

Smart devices are also used to find local services, such as hospitals, schools, libraries and other places within a city or town. People can use apps on their devices to find events occurring within the city or town, and to pay for them.

This woman is paying to use transport in a city.

HELP FOR TOURISTS

Tourists visiting a smart city can use free Wi-Fi to find interesting places to visit, such as museums, parks and play spaces.

Smart Schools

Schools in smart cities and towns use lots of electronic devices connected to the internet to help students and teachers study and work.

Virtual reality headsets help students learn about subjects such as history and science. A virtual reality headset can make the wearer feel like they've gone back in time, or travelled to the Moon!

Virtual reality headsets make you feel like you are somewhere else.

Computers, tablets, touch screens and other electronic devices are also used in smart classrooms. Students may work together in small groups, or on their own, using these devices.

CLASSES AT HOME

Some students study in a "virtual classroom". Students can attend lessons each day from their home, using a computer. They don't have to live in a city or town to go to a virtual school.

Traffic and Technology

Smart cities and towns use technology to keep traffic moving around the roads, especially at busy times such as mornings and evenings, when people travel to and from work and school. Cameras and other sensors gather information on traffic numbers and speed, and control how often traffic lights change.

Traffic cameras also help keep roads safe. They can detect drivers speeding, or driving through red lights. They can also detect people using smartphones while driving. Sensors at pedestrian crossings detect how many people are waiting to cross busy roads. This information can be used to help make crossing roads faster and safer.

Cameras monitor traffic and feed the information back to a control centre.

Sometimes traffic jams happen in large cities and towns. Then roads can become unsafe and accidents can occur. In smart cities and towns, cameras and sensors send information about traffic movement to a control centre. When there is an accident, help can be sent to the scene immediately.

Drivers can also use an app on their smart device to find traffic information, or use a **navigation system** in their car, to avoid traffic jams and find the best way to get to where they want to go.

This control centre is managing traffic on busy roads.

SMART PARKING AND CHARGING

Smart parking sensors are used in smart cities and towns. Drivers can use an app to find places where parking spaces are available, or places where electric vehicles may be charged.

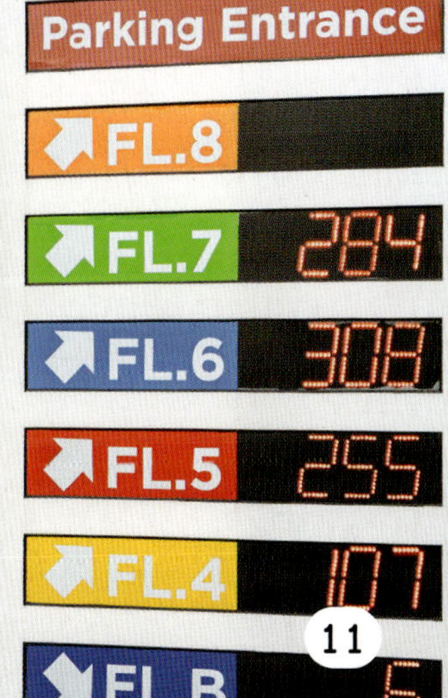

This board shows how many parking places are available on each floor, or level, of the car park.

Public Transport

Public transport plays an important part in taking people from place to place in smart cities and towns. Many people use public transport to commute, or travel, to work each day from areas beyond a city or town. When many people commute using public transport, it helps save energy and makes travel more sustainable.

Driverless trains don't have a driver. Instead, the trains are managed by operators at a control centre. Information is gathered from cameras and sensors inside and outside the trains to help keep the trains moving safely and on time.

IN CASE OF EMERGENCY!

Most driverless trains still have a train operator on board, who can take over in case of emergency.

This driverless train operates in Sydney, Australia.

Driverless buses are used in some cities. These buses use sensors, cameras and computers to travel safely on the roads. Driverless buses still have a driver on board, however! The drivers can override the automatic driving system and take over if needed.

People using public transport in smart cities and towns can access an app on their phone to find a train station or a bus or tram stop. They can also use an app to pay for their trip.

This driverless bus in France is operated from a control centre.

Smart Play Spaces and Parks

Technology is used in smart cities to encourage people to enjoy playgrounds, parks and other green spaces. These places help people stay healthy by exercising, having fun and relaxing.

Playground equipment is designed to encourage children to use it and be active. Smart playgrounds have small points on the play equipment that connect with apps on smartphones. Children and adults can play games that use both a phone and the equipment. Users get lots of exercise as they move around the play equipment, playing the games.

Smart playgrounds with things like interactive touch lights are exciting places to play.

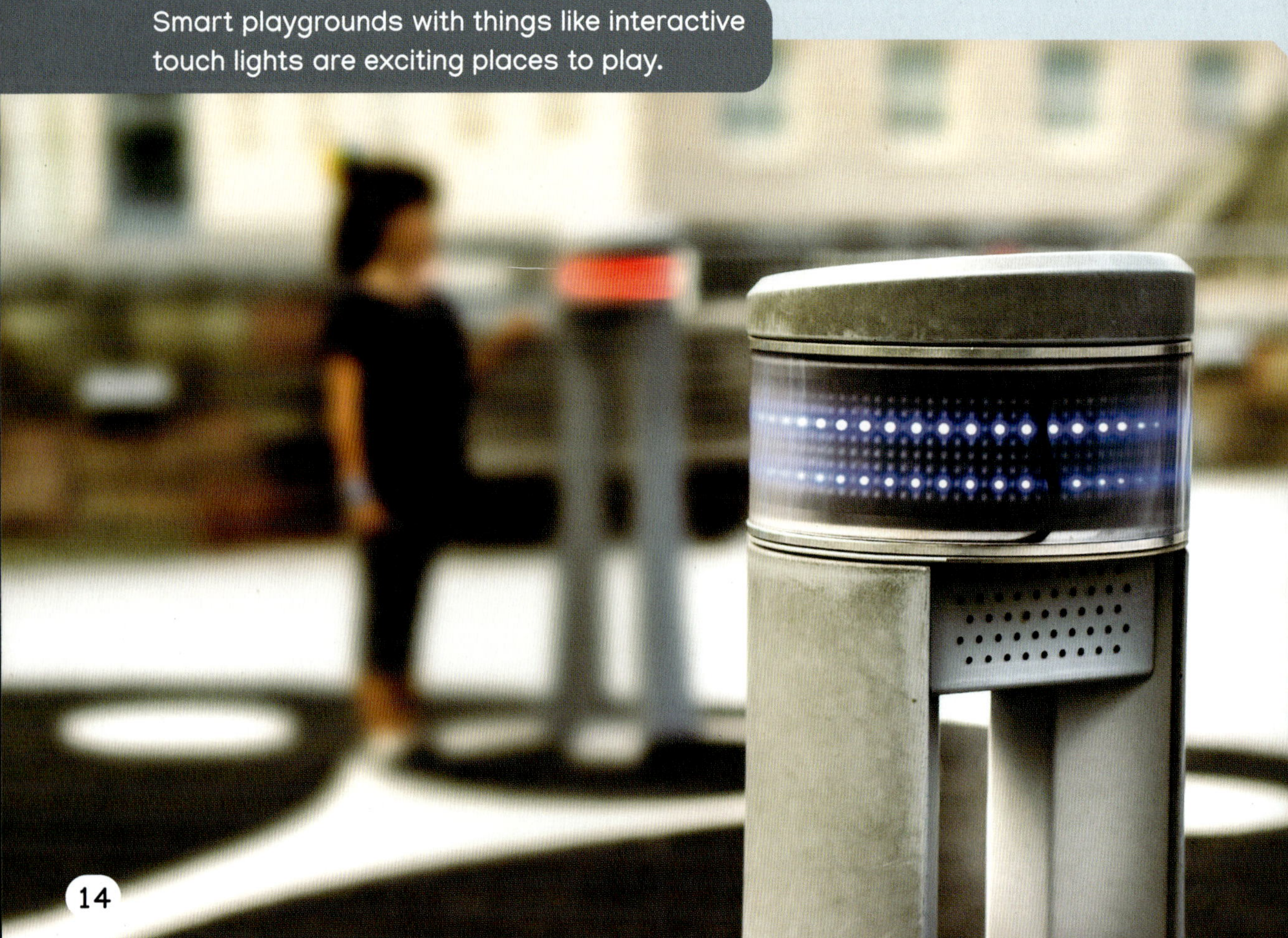

Smart cities and towns use technology to help care for plants in **urban forests**.

The thousands of buildings in cities and towns can become very hot when sunlight heats large areas of concrete and other building materials. This makes the environment in cities and towns warmer than in other places.

Trees in urban forests help keep cities cooler. Small sensors on and around some trees in smart cities collect information about the trees, such as whether they need watering. Computer programs can store the information collected to help keep trees healthy.

Sensors can help to maintain the health of complex urban forests.

LONDON

The city of London, in England, is one of the world's smartest cities. It also has one of the world's largest urban forests. There are more than 8 million trees in the city of London.

Electricity, Water and Waste

Electricity for Cities and Towns

Cities and towns use huge amounts of electricity. An **electricity grid** carries power to homes, offices and other places where it is needed.

Sensors, controllers and **meters** monitor and control electricity delivery in smart cities and towns. This helps stop waste and conserves energy.

Transmission towers move electricity from power stations through the electricity grid to buildings in towns and cities.

Water Needs

Everyone needs water! Modern cities use huge amounts of water in homes, office buildings, and parks and gardens. Smart water meters transmit information to water companies, helping to detect water leaks and monitor water use. This helps to save water.

smart water meter

POWERED BY THE SUN

Solar panels help to power many smart cities and towns today. These panels can be attached to buildings or set up on the ground.

Smart meters send information to electricity companies about the amount of electricity each set of solar panels is generating and providing to the electricity grid.

Waste Management

Cities and towns create lots of waste. In some smart cities and towns, bins are fitted with sensors that send a message to garbage collectors when the bins are almost full. Garbage collectors empty the bins when they receive this information.

Other cities use a system of underground pipes to dispose of waste. Waste is placed in chutes above the ground. Sensors send a message to a collection station when chutes are full. Then fans push air into the pipes and waste is sucked through them to collection points.

Smart waste removal saves time, money and energy because fewer garbage trucks need to travel around cities and towns.

These bins have sensors that alert garbage collectors when they are full.

CASE STUDY

Melbourne

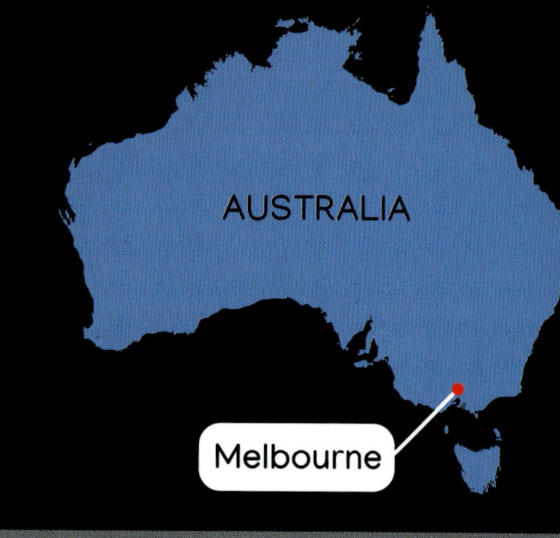

Melbourne is the capital city of the state of Victoria, in Australia.

Melbourne is a large Australian city. **Smart technology** is used by the local government to help make it a smart city.

Royal Park is a large park near the centre of Melbourne. Sensors are used to help make the park an even better place to visit. Sensors capture information such as which areas of the park are used the most.

In other parts of Melbourne, the city runs tests using sensors to help the environment. In some places, rain gardens have been built from recycled materials to help filter rainwater from the streets and provide a place for plants to grow. Sensors collect information on how well the gardens work.

Some parks around Melbourne are using smart technology, such as sensors, to help make it an even more sustainable and enjoyable city.

COUNTING PEOPLE

Many pedestrian-counting sensors are used throughout the City of Melbourne. Everyone can look at the sensor sites on a local government website and read the information gathered. The information helps the government learn how people move about the city.

Melbourne has many contemporary buildings and spaces, such as Federation Square, also known as "Fed Square". Fed Square is a venue that includes a large open square, a museum, art exhibitions and places to eat. Modern technology is used at Fed Square to benefit all who visit it.

Fed Square has a high-tech design that saves energy and makes it more sustainable. A giant maze of specially shaped concrete walls under Fed Square helps to heat and cool parts of it.

A large screen faces the square, allowing visitors to watch sport and other events. Free Wi-Fi is available here for visitors, too.

People can gather in Fed Square to watch sport and other events together on the large screen.

CASE STUDY

Singapore

Singapore City is the capital ci of Singapore.

The island of Singapore is home to one of the smartest cities in the world.

Singapore has a special Smart Nation project that aims to help every citizen benefit from smart technology and the Internet of Things, so that nobody is left behind in the "digital world". People can use a free government app that helps them do things such as find government health services and information, and learn what is happening in the community. They can even use it to set up a monitoring system for elderly family members.

Singapore is one of the world's smartest cities.

BUILDING A SMART NATION

Everyone in Singapore can take part in helping to make their country one of the smartest in the world. A Smart Nation Builder truck stops at different cities and towns, so everyone can try smart technology and offer suggestions on how to improve Singapore's smart cities and towns.

Gardens by the Bay is a group of public gardens on Singapore's waterfront. People can visit the gardens and see plants in two giant conservatories, or greenhouses: the Flower Dome and the Cloud Forest. The conservatories have huge shade sails, operated by sensors, that can be rolled out to stop the precious plants inside from getting too hot.

Visitors to the garden can also see enormous "supertrees", which are vertical gardens – gardens grown up a vertical surface. Some of the supertrees have solar cells on them, which collect solar energy and provide sustainable light at night.

These "supertrees" rise up to 50 metres above the ground.

CASE STUDY

New York City

New York City is the largest city in the United States of America. The government of New York City is using technology to improve services such as public transport and city security, to make it a better place to live.

A street lighting program is replacing old lights with new **LED lights**, which use less energy and make the city more sustainable. Some light poles have cameras and other sensors on them to help keep people safe and count traffic.

Old pay phones are being replaced with free Wi-Fi **kiosks**. People use the kiosks to make calls and charge their phones.

New York City is the capital city of the state of New York, in the United States.

Protecting the City

New York City's government runs a "Cyber Academy", where employees from government services are trained in **cybersecurity**. Here, they learn how to protect those services from computer hackers, viruses and other crimes on the internet.

People can connect to free Wi-Fi at this kiosk.

New York City's subway is one of the largest and oldest train services in the world. There are 472 stations in the city. Modern upgrades are being made to the subway to make it smarter. Its signalling system, which controls how trains move around the railway tracks, is being replaced with a new automatic system.

The "communications-based train control system" is a computerised system. Pieces of equipment on the train and along train tracks communicate with each other using radio signals as the train passes. This tells the control system where each train is. Information is also sent to a control centre where trains are monitored.

Smart technology helps this train to be safer and more reliable.

Technology Everywhere

The same technology used in large cities and towns is also benefiting smaller towns and remote places around the world. Thousands of internet satellites orbit Earth. Remote towns, schools, farms and homes are now able to use high-speed satellite internet connections to contact the rest of the world. This makes it easier for people to study, work and connect with friends and family, just as people do in larger cities and towns.

Satellites orbit Earth, communicating with devices connected to the internet below.

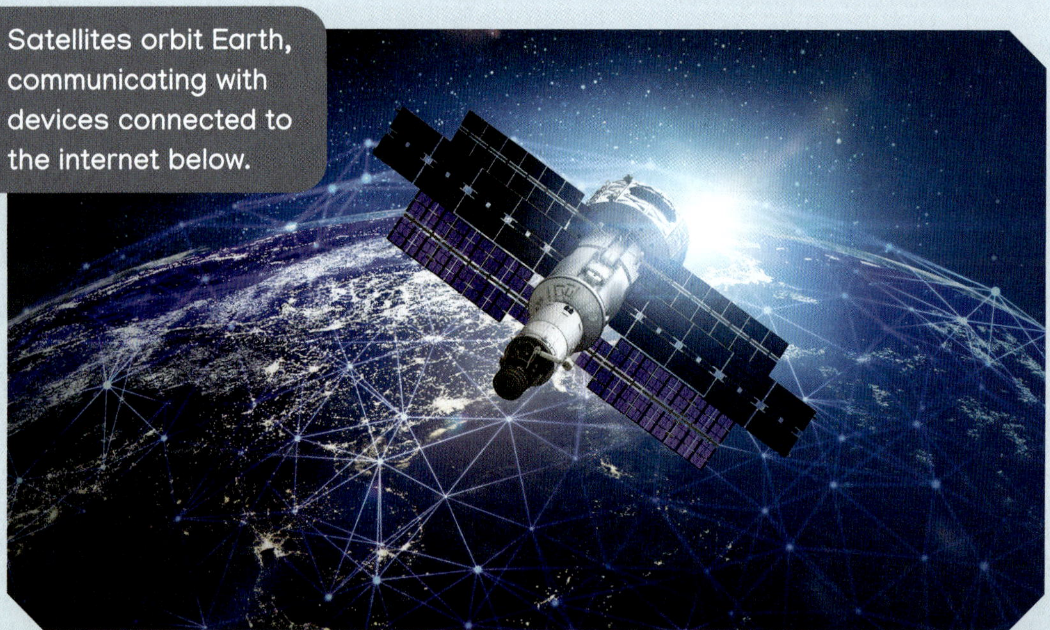

POWER IN THE OUTBACK

The remote Australian town of Coober Pedy has its own mini power grid, which uses wind and solar power. Citizens can check a website to see how much of their power is coming from renewable energy each day.

Many smaller towns are also working towards being smart places to live. Like larger cities, they are using the Internet of Things to help make life better for all.

Free Wi-Fi is available in most smaller towns at places such as libraries, local government offices and bus and train stations. Visitors to tourist towns can download apps to their mobile phones to find information on things to do in town.

People can charge their smart devices and use free Wi-Fi at this solar-powered park bench.

Smartphones allow tourists to easily and quickly find information about a city.

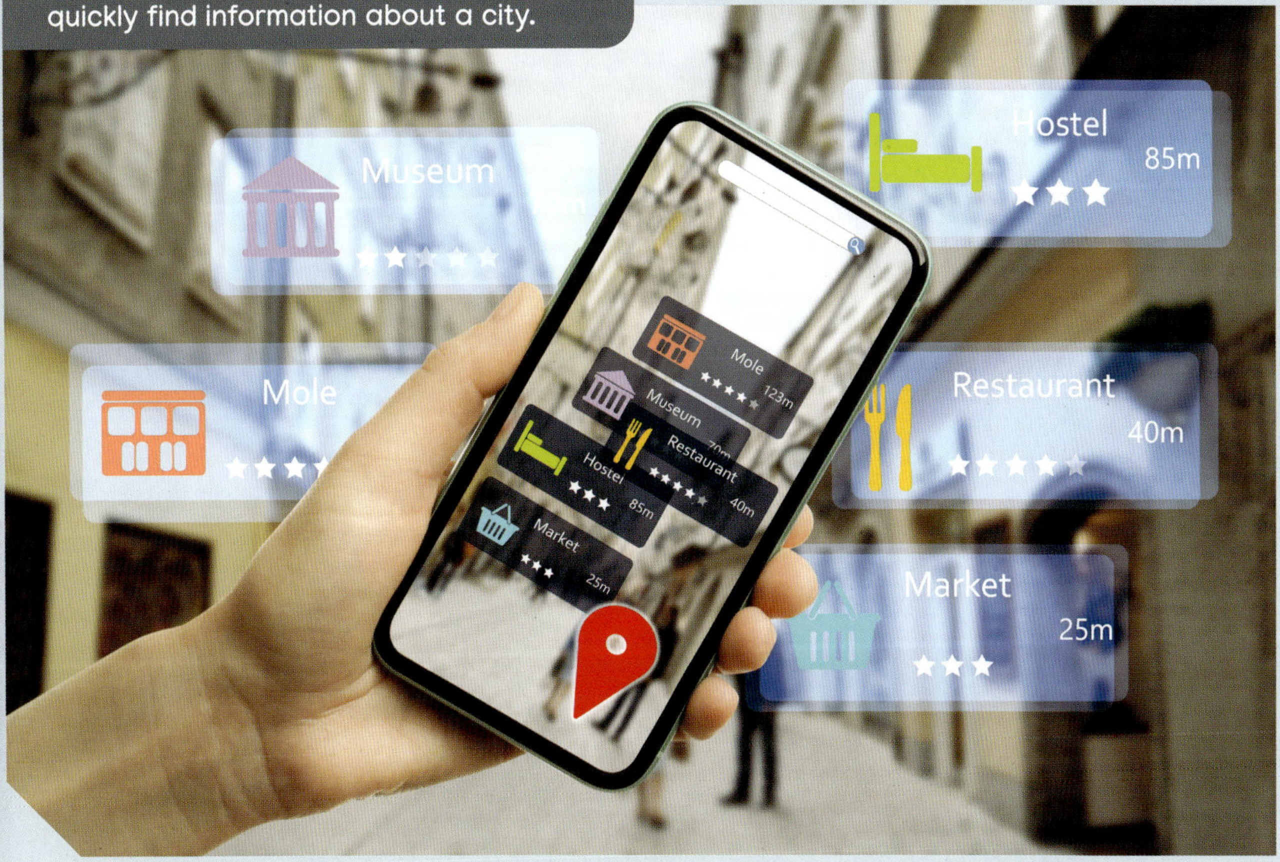

A Fantastic Future!

Many places around the world are becoming smart cities and towns. The lives of people who live, work and play in these places are improving in many different ways.

Smart technology makes it much easier for people to communicate with each other than in the past. Travelling from place to place around smart cities and towns is easier and uses less energy than ever before.

Public transport is also becoming faster and safer with the use of modern technology.

This robot was designed to help rescue people from natural disasters.

ROBOT HELP

Digital technology can be used to do many jobs that assist humans. Robots can do jobs such as helping to build new buildings in cities and towns. Perhaps robots will do many more jobs in smart cities and towns of the future!

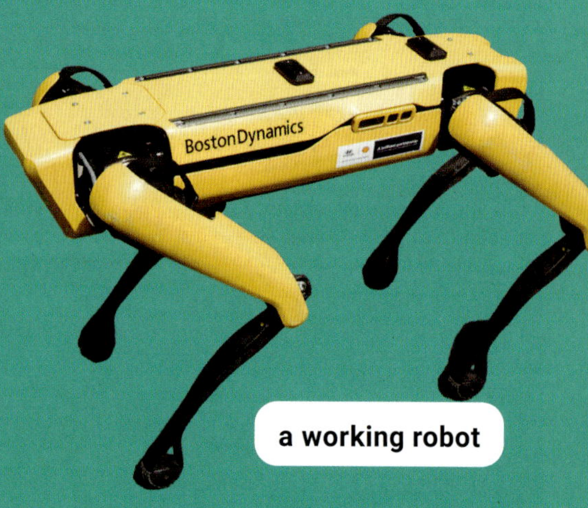

a working robot

Smart cities and towns of the world are becoming better, more environmentally friendly places to live, too. New buildings with smart, energy-saving features are helping cities become more sustainable. Public parks and other public places are managed better with the help of information gathered using IoT devices, such as sensors. Services such as electricity, water and waste management have all become more efficient in smart cities, too.

New digital technology is being invented every year, which will make smart cities and towns even more exciting places to live in or visit!

a vertical garden, monitored by sensors

a modern smart city in China

a Singapore police robot

Smart Cities and Towns – Is It All Good?

A Debate

Modern technology such as smart devices and the Internet of Things have brought many benefits to cities and towns around the world, and improved the way millions of people live, work and play.

However, there can be some problems with this technology, too.

Sharni - Agree

I believe that smart cities and towns are good for everyone.

Smart technology helps people live a more sustainable life.

Information collected by smart meters helps people and governments learn more about things like water and energy use, and manage these services better.

Also, smart technology keeps people safe. Cameras and other sensors help prevent traffic jams, stop drivers from speeding and get help if there are traffic accidents.

Smart devices send and receive information quickly, and allow us to do things such as turn on home security cameras using our smartphone.

Smart cities and towns are good places to live. They help us to connect with each other and do important daily tasks faster, and they also help us to be safer.

Home security systems can be operated via smartphones.

Jayden - Disagree

I believe there are problems with the use of so many electronic devices in smart cities and towns.

Digital technology needs electricity to work. If the power goes off, we cannot use it. We cannot charge batteries either. This means cities and towns cannot use electronic machines once batteries run down. They must find backup power sources, such as **generators**, to produce electricity.

Thousands of cameras in cities record people as they move about. Everyone loses their right to privacy! People's personal information, such as things they like to buy, can be collected using the Internet of Things. It may be stored by governments or businesses. Sometimes, hackers steal people's information despite the technology we use to protect it!

We need to be careful when we use digital technology and do more to protect our personal information.

Generators provide power when the electricity grid fails.

Glossary

authorities (*noun*)	people or organisations with political or administrative power
citizens (*noun*)	people who live in a particular place
cybersecurity (*noun*)	the application of technologies and processes to protect systems, networks, programs, electronic devices and data
digital technology (*noun*)	electronic machines such as computers and mobile phones, and the programs that run on them, or make them work
electricity grid (*noun*)	the connected machines and wires that take electricity from where it is made to places where it is used
generators (*noun*)	machines that can make electricity, and that often run on petrol or diesel fuel
kiosks (*noun*)	small booths, or buildings with open sides, that contain services such as internet access or house small shops
LED lights (*noun*)	a modern kind of light that uses very little energy; LED stands for light-emitting diode
liveable (*adjective*)	suitable or acceptable for living in
meters (*noun*)	machines that measure, or measure and record, things such as water, electricity, distance or speed
navigation system (*noun*)	a system in a car or other vehicle that helps people find where to go when they are driving
networks (*noun*)	groups of computers or other technological devices connected to each other
sensors (*noun*)	machines that can sense and collect information about movement, light and other things
services (*noun*)	things that are provided by the government or businesses, such as public transport, electricity and water
smart devices (*noun*)	electronic machines such as phones or computers that can connect to the internet to send and receive information
smart technology (*noun*)	sensors, monitors, cameras and other electronic machines that can connect to the internet and send and receive information
solar panels (*noun*)	panels that convert sunlight into energy

sustainable (*adjective*)	able to be used now and also in the future, without harming the environment or using something that cannot be replaced
urban forests (*noun*)	trees and other plants that grow in a city or town
virtual reality headsets (*noun*)	devices you wear over your eyes that shows you something that looks real but isn't

Index